Edilson Ramos de Lima

Bomba Carcará - Social technology for handling rainwater

Edilson Ramos de Lima

Bomba Carcará - Social technology for handling rainwater

Efficient social technology that prevents the emergence of diseases caused by improper handling of rainwater

ScienciaScripts

Imprint

Any brand names and product names mentioned in this book are subject to trademark, brand or patent protection and are trademarks or registered trademarks of their respective holders. The use of brand names, product names, common names, trade names, product descriptions etc. even without a particular marking in this work is in no way to be construed to mean that such names may be regarded as unrestricted in respect of trademark and brand protection legislation and could thus be used by anyone.

Cover image: www.ingimage.com

This book is a translation from the original published under ISBN 978-613-9-64021-8.

Publisher:
Sciencia Scripts
is a trademark of
Dodo Books Indian Ocean Ltd. and OmniScriptum S.R.L publishing group

120 High Road, East Finchley, London, N2 9ED, United Kingdom
Str. Armeneasca 28/1, office 1, Chisinau MD-2012, Republic of Moldova, Europe
Printed at: see last page
ISBN: 978-620-7-72048-4

Index

CHAPTER 1	4
CHAPTER 2	6
CHAPTER 3	10
CHAPTER 4	15
CHAPTER 5	18
CHAPTER 6	21
CHAPTER 7	24
CHAPTER 8	33
CHAPTER 9	35

Summary

The "School Cisterns" project, which is being implemented in the semi-arid Brazilian state of Alagoas, is an extremely important tool for providing access to water for school communities. It is an initiative of the Brazilian government and is supported by the Spanish Agency for International Development Cooperation (AECID). Harvesting rainwater guarantees clean water by natural distillation thanks to the water's hydrological cycle, whose chemical characteristics are better in rural areas, as well as some care in capturing, storing and withdrawing it using hand pumps.

The hand pumps are made using social technology and are an instrument of fundamental importance for removing water from the school cisterns, avoiding contamination when removing water for human consumption and ensuring that this water does not come into direct contact with people, since it is the last process of collecting the water. The Carcará 1 and 2 pumps are based on the principle of positive displacement and require little effort. Their benefits lie in their easy construction, high performance and reproducibility, as well as low contamination of the water stored in the cisterns. The construction we have carried out in the laboratory is purely didactic in nature to teach the operation of the Carcará 2 pump, in which the path of the water and the arrangement of its elements will be observed.

Keywords: water, rain, cistern, hand pump, catchment, didactic.

CHAPTER 1

Introduction

The right to water and sanitation (United Nations Resolution 64/292) has been established as a prevalent need throughout the world, but this should include some minimum conditions that guarantee its quality, such as its colour, smell and taste. It should also be accessible for a sufficient amount of survival of approximately 20 litres per day per person, with access less than 1 kilometre away and a collection time of between 5 and 30 minutes (HOWARD 2003). Despite this, there are still many people who do not have access to water, although it is certain that by 2010, one of the Millennium Development Goals of halving the proportion of people without access to drinking water by 2015 will have been met, it is important to know that by the end of 2010 at least 11% of the world's population will not have access to drinking water. In Latin America and the Caribbean, the data is more encouraging, with 94% of the population having access to drinking water, but the differences between urban and rural areas are still great. In Brazil, the data shows that in the last 10 years, access to water in rural areas has increased from 77 to 85%, while in urban areas it has increased from 98 to 100% (UNICEF / WHO 2012).

Over the last ten years, Brazil has promoted social programmes such as the "One Million Cisterns Programme" (PIMC) run by the "Brazilian Semi-Arid Articulation" (ASA), a network of organisations that manage policies for coexistence with the Brazilian semi-arid region. The PI MC programme is based on the use of rainwater in Brazil's semi-arid region, one of the wettest in the world, where climatic conditions, 3 to 5 months of rain, average rainfall of 750mm/year, ranging from 250,800mm/year and significant evapotranspiration (evaporation of 3000mm/year. Malvezzi, 2007) mean that rainwater harvesting is one of the simplest and lowest-cost ways of alleviating the region's water deficit.

The "Cisterns for Schools" programme is being carried out by the Brazilian government to meet the social development needs of the population in Brazil's semi-arid region. The "Consortium for the Development of the Ipanema Region" in collaboration with the "Cooperation Fund for Water and Reparation" (CONDRI), the Spanish Cooperation instrument (FCAS) and the Polytechnic University of Madrid (UPM) are carrying out the project "3ª Agua. Dilui para Educar" to build 52,000 litre cisterns for 108 schools in the semi-arid region.

CHAPTER 2

Third Water Cistern Programme

The state of Alagoas, whose territorial area is approximately 27,779.34 km^2 and according to data from the last Demographic Census carried out by the Brazilian Institute of Geography and Statistics (IBGE) in 2010, has a population of around 3,120,494 inhabitants, is characterised by high vulnerability to drought. This is due to the fact that 42.80% of its territory is in the so-called sertanejo region, which has peculiar characteristics (water deficit and high evapotranspiration), making it different from other regions in Brazil. In this sense, the construction of plate cisterns in the Sertão de Alagoas with a storage capacity of 52,000 litres of rainwater for the consumption of pupils and maintenance of rural schools, in the context of the adoption of social technologies, represents an investment in proven, effective and low-cost alternatives for the universalisation of water shortages in 108 schools in 13 municipalities.

The "Water in Schools" initiative, also known as the Third Water Cisterns Programme or "Water for Education", is part of the BRA-007-B Cisterns Programme signed in 2009 between the Spanish Cooperation Agency (AECID), the Sustainable

Brazilian Environmental Institute (IABS) and the Ministry of Social Development (MDS); whose main objective is to contribute to social transformation by promoting the preservation, access, management and valorisation of water as an essential right to life and citizenship, broadening the understanding and practice of sustainable coexistence and solidarity with the Brazilian semi-arid region. In addition to third-water cisterns, it includes other lines of action, such as the construction of household cisterns (first water) and cisterns for small-scale production activities in families (second water).

In the Sertão region of Alagoas, the Consortium for the Development of the Ipanema Region (CONDRI), a public, non-profit organisation, is responsible for implementing the "Water for Education" project, which ended in March 2014, as well as the First and Second Water Cistern Project. Also an active participant is the Polytechnic University of Madrid, which, through the Centre for Innovation in Technology for Development (itdUPM) and several of its associated groups, is carrying out an impact assessment consisting of: a technical assessment of the construction of the cisterns; an assessment of the management model employed; an analysis of the quality of the water supplied and an impact assessment of the improvements in the living conditions of the school community due to the implementation of the school cisterns.

Figure 1. Rainwater harvesting cistern at a school in Major Isidoro, Alagoas.

As far as construction is concerned, it should be borne in mind that the cistern must be buried from half to two thirds of its height. It consists of cement slabs measuring 50 by 60 centimetres and 4 to 5 centimetres thick, curved according to the projected radius of the cistern wall. The slabs are made on site in wooden or iron moulds. The cistern wall is raised with these cement slabs, starting from the already cemented floor. To prevent the wall from collapsing during construction, the slabs are glued together with cement mortar and the cement paste is allowed to dry for 8 hours. Galvanised steel wire is then wrapped around the outside of the cistern wall. Finally, the roof is built with other pre-moulded slabs in a triangular shape, using a weaker mortar mix, and these are placed on top of reinforced concrete beams, then the external roof is plastered. After the cement has dried, the entire cistern is painted with whitewash. The construction also includes excavation, renovation of the roof, installation of the Carcará 2 pump to remove the water (Figure 2), installation of the gutters and spouts and finally identification with a numbered plaque.

Figure 2 - Installation of the first Carcará 2 pump in a school in Major Isidoro, Alagoas (October/2012).

The cistern is considered to have been built (completed) only after it has been identified, georeferenced, photographed and the school community, through the school headmaster, has signed a document declaring that the cistern has been received in perfect condition.

CHAPTER 3

The path of water in the Programme's schools. Quality and funding

It makes sense to use the term "water path" to refer to the water's journey from its source to the point of direct use in schools, with the intention of highlighting and collecting all the possible causes and/or routes of contamination from the water's arrival at the cistern to its consumption. In this way, it is possible to differentiate between several critical points of pollution: pollution at the source, at the cistern, when the water is extracted from the cistern and along the route from extraction to use.

The risk of water pollution depends primarily on where it comes from. Some of the schools receive piped water from the São Francisco River in Pão de Açúcar via the Alagoas Sanitation Company (CASAL), with chlorination treatment. However, the vast majority are supplied by water tankers, which can be from the Civil Defence (under the Operação pipa and Agua é vida programmes) or the municipality, and although the water in both cases usually also comes from CASAL, the risk of pollution is increased by the hygienic conditions of the water tanker, as well as the decrease in chlorine during transport from the treatment plant to the school in question. However, it should be noted that due to the outbreak of diarrhoea that occurred in more than 52 municipalities in the state between June and August 2013, resulting in 52 deaths, the sanitary controls of the water tankers have been reinforced, and even in some of the municipalities, residual chlorine monitoring has been set up.

Generally speaking, in rural areas, far from industrialised zones and therefore without atmospheric pollution, rainwater is of acceptable quality for human consumption due to the natural distillation process of the hydrological cycle. Even so, rainwater is very susceptible to contamination from the catchment area to the cistern, which is why CONDRI's Third Water Cisterns programme includes repairing the roofs of each school; re-roofing, washing the tiles and disinfecting the entire roof, as well as lining it with plastic sheeting to ensure better catchment and control of diseases caused by bats and sparrows. Even so, the disposal of the first waters of each rainfall is fundamental to guaranteeing the quality of the water, and in the event that they are carried out manually, a procedure must be defined to ensure good management. Likewise, the maintenance of pipes and spouts and the annual washing of the roof are essential for this.

The risk of contamination in the cistern itself can be reduced by adopting preventive measures. Firstly, the cistern should not be located near large trees, corrals, culverts, septic tanks and rubbish dumps, in order to avoid infiltration that could occur through cracks or gaps in the cistern. The cistern's outlet must be protected from the entry of small animals and the rainwater inlet must be designed so that there is no turbulence, so as not to remove the silt settled in the soil from the cistern. Excess material or a lack of washing can also cause the cement to detach from the cistern water.

The Carcará II pump plays a fundamental role in preventing contamination inside the cistern. During the UPM team's first contact (which included a visit to the 58 schools in the programme), it was found that more than 90% of the schools that already had a cistern before the project (many of which were damaged or poorly maintained) draw their water using the traditional method, with a bucket tied to a rope (Figure 3). This can lead to contamination of all the water in the cistern. The Carcará II pump reduces the risk of pollution by allowing the cistern to be kept closed at all times, even when it is being collected.

Figure 3: Extracting water from an old cistern with a bucket tied to a rope at a school in Canapi, Alagoas.

The path of the water from the withdrawal of the water to its use. The use of properly washed containers categorised by use, with the intention of not using the same storage container for washing as for consumption, as they do not require the same hygienic conditions, should be carried out as far as possible. Many of the schools in the programme currently use ceramic filters to reduce the pollution of drinking water, and more than 70% say that they normally use chlorine provided by the Ministry of Health, although they also often claim that it is not supplied in sufficient quantities. If the preventive barriers mentioned above are not carried out successfully, with studies and monitoring to prove it, it is essential to establish corrective measures, such as chlorination, to ensure that the bacteriological quality of the water is fit for consumption.

Thanks to a small preliminary study carried out by CONDRI, 8 litres/day have been identified as the average consumption per pupil, but more details are still to be confirmed in forthcoming studies. Taking into account the irregularity of annual rainfall and the number of pupils per school (ranging from 12 to 429 pupils), it is possible to predict that exclusive rainwater supply will not be guaranteed in some of the schools, in many cases having to resort to traditional water sources.

CHAPTER 4

Carcará pumps 1 and 2

The "Carcará 1 and 2" pumps are part of the development of social technologies that have been adapted to the badge cistern programme thanks to their low cost and easy construction in a way that, just as in the construction of badge cisterns, the population participates in their development with simple tools that make a social commitment exist to achieve a very common goal.

The use of hand pumps guarantees a low level of contamination in the extraction of the water stored in the cisterns, which, together with good care and cleaning of the reception area and the distribution pipes, include the recommendations that the WHO gives for the reduction of health risks for rainwater (WHO 2012). In this sense, the "Carcará 1 and 2" pumps are a great tool for carrying this out.

A pump was built in the laboratory of the School of Industrial Design and Creation (ETSIDI) of the Polytechnic University of Madrid (UPM)

Carcass 2 with methacrylate for a didactic purpose so that the internal elements and the flow of water can be observed. They have undergone some changes compared to the original version for scales available in the laboratory, but we considered even this way that their operation was not affected.

In Alagoas, the Carcará 1 and 2 pumps are manufactured locally in a workshop in São José da Tapera, a municipality that is also a beneficiary of the First and Third Water Programmes (Photo 4 and 5). With a team of six technicians and more than 10,000 Carcará 1 pumps made for the Home Cisterns Programme in recent years, the Tapera team is in charge of making the 108 Carcará 2 pumps for the school cisterns.

 The Carcará 2 pump is intended to be more practical, as well as more economical compared to its predecessor (the cost of which is around R$100 compared to R$130 for the Carcará 1 pump). Past experience has shown that the installation of the pump is fundamental to its durability, as some of the pumps installed have lost their lower part due to not being glued together, becoming inoperative and not being serviced. Consequently, the Carcará 2 pump is being glued and modifications are being studied to improve the fixing.

Figure 4 - Assembly of Carcará Pump 2.

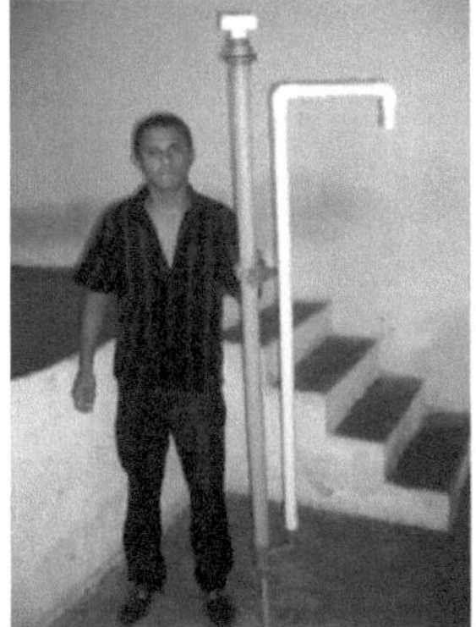

Figure 5 - Complete assembly of Carcará Pump 2.

CHAPTER 5

Carcará pump 1

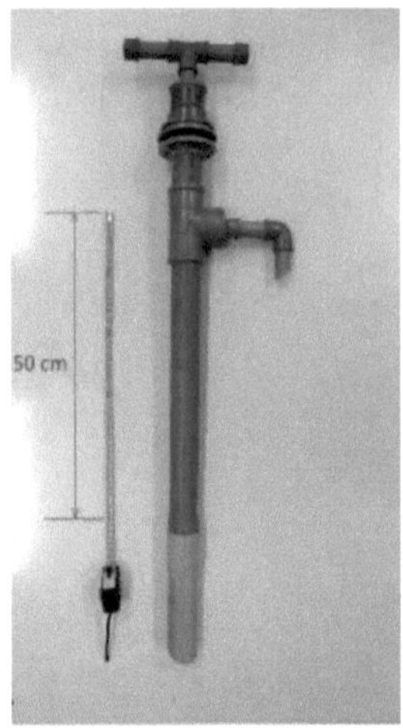

Figure 6 Carcará pump 1.

5.1. *Materials*

- PVC pipes of 50 PVC diameters
- 32 PVC pipe
- Tube 25 PVC

- 20 PVC pipe
- Tapered reduction 50x32 PVC
- Tapered reduction 32x20 PVC
- Codo 32 Hembra - Hembra
- Te 50x50x32 90 grados soldable - roscable
- Te 20 90 grams weldable
- Tapón 32
- Tapón 25

- Tapón 20
- Casing 32x25
- 32x1" mixed thread cuff - enclose
- Tapered reduction 40x25 PVC
- Casing 40x32
- D 30mm crystal ball

5.2. *Tools*

- Grinding motor
- Taladro
- Drill No. 8
- Martillo
- Pliers
- Sierra de mano
- Threader
- Metro
- Industrial mechanic
- Grinder
- Lima
- PVC grip
- Mascarilla
- Protective gowns

5.3. *Operation*

The principle of operation of this pump is that of a positive displacement volumetric pump in whose interior there are two check valves that only they allow to pass the water up having formed by two marbles or marbles, one in the inner tube joined in a conical reduction of 50x32 and another in the mobile tube that is joined to this. When you raise the movable tube or piston, negative pressure is exerted on the lower valve, opening it and allowing the water to pass through.

in its interior being filled the existing space, in return he/she surrounds the valve the piston that displaces the water upwards making that the exit hole comes out in it You. Then, when the movable tube or piston is lowered, the pressure increases in the water inside, which causes the lower valve to close and the piston valve to open, allowing the water to pass through the holes specialised in the piston, thus following Archimedes' principle, the volume of water displaced by the piston will exit into the outlet hole in it, which completes a pumping cycle.

CHAPTER 6

Carcará 2 pump

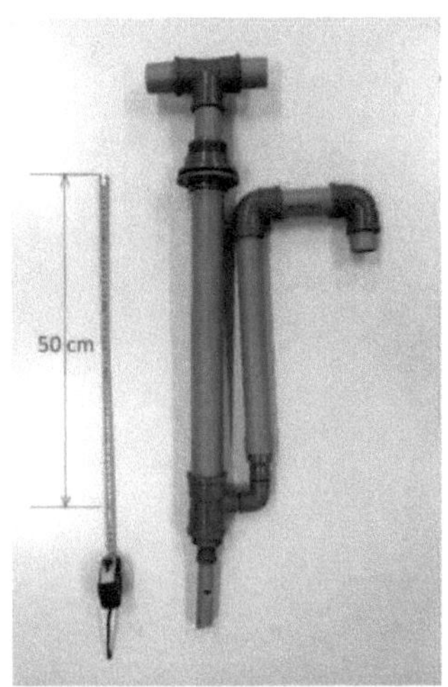

Figure 7. Carcará pump 2.

6.1. *Materials*

- Tapered reduction 50x25 PVC
- Tapered reduction 40x25 PVC

- Codo 40 Hembra - Hembra PVC
- Codo 25 Hembra - Hembra PVC
- Tee 50x50x25 90 grams PVC
- Te 40 90 grit PVC
- Tapón 40 PVC
- 50 PVC pipe
- 40 PVC pipe
- Tube 25 PVC

- D 30mm crystal ball
- Wooden palette

1.2. Tools

- Sierra de mano
- Clavo
- Martillo
- Lima
- Cúter
- Metro
- Labelling
- Mechero
- PVC grip

1.3. Operation

The operation is similar to that of the Carcará 1 pump, being based on the positive displacement principle. It also has two check valves that only allow water to pass through. When the piston rises, negative pressure is exerted on the lower valve, which raises the marble that passes water inside, filling the existing space and, in turn, also on the 40x25 conical reduction, which closes this valve and prevents the water in its upper part from flowing down. Then, when the piston is lowered, the pressure increases in the water inside, causing the valve in the 50x25 conical reduction to close and the valve in the 40x25 reduction to open upwards, letting the water flow outwards and completing the pumping cycle.

CHAPTER 7

Case study. Carcará 2 methacrylate pump

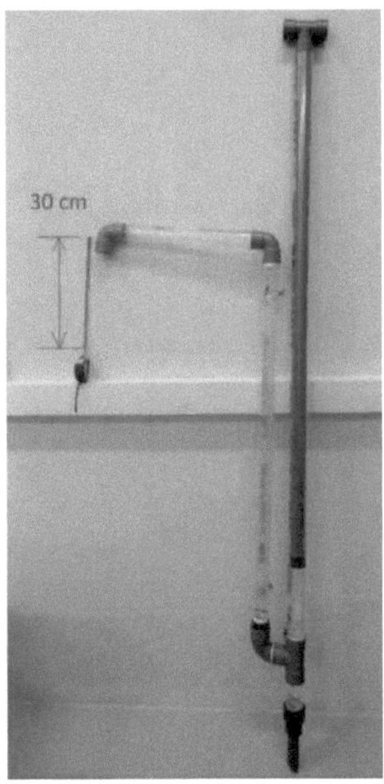

Figure 8. Carcara 2 methacrylate pump

At ETSIDI's "Hidráulica Aplicada al Desarrollo" laboratory, a model of the Carcará 2 pump has been built. We used the following materials and tools to build it:

- Methacrylate tube 50x46
- Methacrylate tube 40x36

- Tapered reduction 50x25 PVC
- Tapered reduction 63x40 PVC
- Reducing socket 40x32 PVC

- Codo 40 Hembra - Hembra PVC
- Codo 50 Male - Hembra PVC
- Te 50 90 grit PVC
- Te 40 90 grit PVC
- 50 PVC cuff
- Tapón 40 PVC
- 40 PVC pipe
- Tube 25 PVC
- D 30mm crystal ball
- D 25mm crystal ball
- Plastic paio
- Wooden palillo
- PVC grip
- Silicone
- Clavo 2.5mm
- Mechero
- Teflon

- ***,1, Construction***

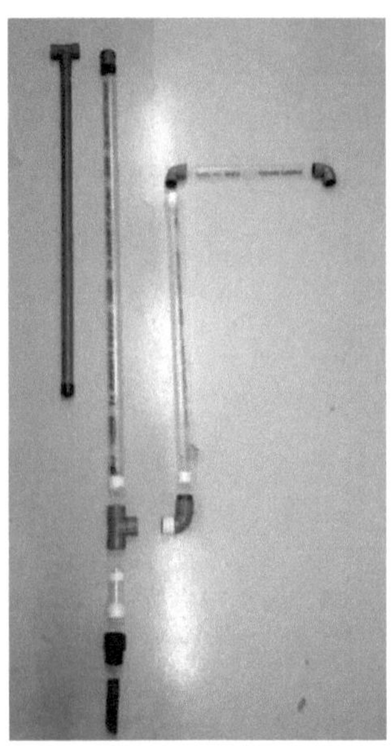

Figure 9. Mounting diagram

The 50x46 methacrylate tube is cut to measure 13 centimetres and 150 centimetres, the 40x36 methacrylate tube to 100 centimetres, 40 centimetres and 10 centimetres, the 40 PVC tube to 130 centimetres and two pieces of 10 centimetres each. To assemble the plunger 40 PVC buffer rods at one end of the PVC pipe 40 is machined into a tomo on the outside of the cap is 50mm in diameter is reduced to 45mm, then fiira the Te 40 90 degrees finally non-collinear at the other end of the pipe and the pieces of 10 centimetres of PVC pipe, 40 are glued together with a Te PVC buffer 40. Cut the 25 centimetre PVC pipe into 13.5 centimetres and make a chamfer and a slit to allow the water to flow through smoothly.

To make the lower valve, we take the 50 cm

methacrylate tube and make 2 diametrically opposite marks 5 cm apart on one of the two edges and we practise sending holes in these marks with a previously hot nail, while ensuring that the plastic stick fits snugly. When we've finished, we put the wooden toothpick into the plastic rod and we insert the group into the holes in the methacrylate tube, we mark where we want to cut the plastic rod with the toothpick trying to measure the diameter of the tube and we cut it finally we mark it with silicone ensuring that there are no losses. Then we put in the 40x32 reduction in the 63x40 conical reduction, we put the marble and we put the 50 methacrylate tube to the part where it has the plastic rod, previously with Teflon on the edge so that this is very fixed and without losses.We repeat the same procedure for the upper valve in the 40 lOOcm methacrylate tube, here we will take the 50x25 conical reduction and the marble will be 25mm.

Figure 10. Valve detail

Figure 11. Top valve detail

Once we had all of the above, we began assembly. We cover the extreme end of the remaining pipes and the male part of the 50 elbow with Teflon. Then we place the 50x46 150cm methacrylate pipe in the Te 50 90 degree PVC and at the colineal end of it we place the other 50x46 13cm methacrylate pipe, at the end of which we previously placed the 25 pipe in the lower valve. The 50 M-H elbow is placed at the required end of the 50 M-H elbow, which is previously placed in the upper valve with the 40 cm methacrylate tube, which continues with the 40 cm elbow, 40 cm methacrylate tube, 40 cm elbow and 40 cm methacrylate tube.

Finally, we attach a 50 PVC screw to the end where we will insert the piston so that it doesn't rest directly on the pipe, we put the piston and the mounting ends on.

If you test it several times and correct the different defects it may have in the assembly, you can check that the operation is the right one.

7.2 Calculations

To calculate the flow of the Carcará 2 didactic pump, they took the following considerations into account:

Piston career: 50 cm.
Discharge tank: 30 litres.

The results were as follows:

Time	s	Flow Q=V/t			
		1/s	m/h^3	1/min	1/h
ti	70	0,43	1,54	25,71	1543
t2	67	0,45	1,61	26,87	1612
Í3	66	0,45	1,64	27,27	1636

	69	0,43	1,57	26,09	1565
t5	62	0,48	1,74	29,03	1742
tó	63	0,48	1,71	28,57	1714
Í7	64	0,47	1,69	28,13	1688
t8	63	0,48	1,71	28,57	1714
t9	64	0,47	1,69	28,13	1688
uncle	64	0,47	1,69	28,13	1688
tilde	70	0,43	1,54	25,71	1543
tl2	65	0,46	1,66	27,69	1662
Qmed		**0,46**	**1,65**	**27,49**	**1649**

1 cycle = 2 rows

No. injection		Cycle/s	Cycles/min	Cycles/L	Cycles/m^3
bi	42	0,60	36	1,40	1400
b2	43	0,64	39	1,43	1433
b3	40	0,61	36	1,33	1333
b4	40	0,58	35	1,33	1333

bs	41	0,66	40	1,37	1367
boo	40	0,63	38	1,33	1333
b?	40	0,63	38	1,33	1333
bs	40	0,63	38	1,33	1333
bç	40	0,63	38	1,33	1333
bio	41	0,64	38	1,37	1367
bn	41	0,59	35	1,37	1367
bi2	41	0,63	38	1,37	1367
Media		**0,62**	**37**	**1,36**	**1358**

7.3 Discussion

In building the pump we had a number of inconveniences that are important to remember. Firstly, the upper and lower valves were made as shown in the drawings contributed from Brazil. In it you can see that the wooden pallet was inside the reduction cone separated at a distance of between 2 and 3 mm (Figure 12). When it was manufactured to these specifications it was observed that the effort required for its operation was considerable and that it also emitted a buzzing sound inside, the results of the first experiments showed a time of approximately 3 minutes to fill the 30-litre tank. He placed the plastic rod inside the tube and it was resolved, waters upwards, from each valve (Figure 13).

Figure 12. Reduction assembly
Figure 13. Tube assembly.

Another important aspect is the relaxation between the piston and the tube in which it slides, as it depends on the suction capacity, less relaxation has more suction. In the pump you can see driving from Brazil, the piston is manufactured in such a way that the socket is introduced from the outside into the 40" tube, while preheating the edge and then its outside is cut so that it enters the 50" tube. When preparing only the measurements previously given for the methacrylate, the only solution we had was to put in a 50 socket and mechanise its outside and try it out until it slid properly.

CHAPTER 8

Conclusions

When analysing the teaching case, we could see that the flow rates are acceptable for a person's daily consumption, as well as its ease of construction and handling. As it is a teaching case, the cost of certain materials is higher than for normal PVC and PB production. It should be emphasised that the modifications made were based on the purely didactic nature of them so that the person could observe the internal components and the flow of water, as well as the benefits since by having the marbles move more freely, the work effort is reduced and the flow is greater.

Finally, to complete the study of this pump, it is advisable that it is fixed to some support and that the operation is carried out under similar conditions to those of work, as well as its possible ability to work at greater depths.

Thanks to the evaluation carried out in Brazil, it was noted that the assembly with paste has to be done on specific connections of the pump. It is essential that they have an appropriate separation so that this guarantees good maintenance and functionality of the pump. It was recommended to use Teflon for these connections to complete their task, without which the flow losses in the pump would be affected to an excessive degree.

CHAPTER 9

Bibliographical references

COHIM, Eduardo. *Preliminary evaluation of a hand pump.* 8º Brazilian Symposium on Rainwater Harvesting and Management. ABCMAC. Campina Grande - PB. 2012.
http://www.bibliotekevirtual.org/simposios/8SBCMA C/8sbcmac-a030.pdf

HOWARD, Guy; BARTRAM, Jamie. *Domestic Water Quantity, Service, Levei and Health.* World Health Organisation (WHO). 2003.
http://www.who.int/water   sanitation health/diseases/WSH03.02.pdf

MEIRA FILHO, Abdon da Silva; BARBOSA DO NASCIMENTO, José Wallace; ANTUNES DE LIMA, Vera Lucia. *Pathologies in rainwater harvesting systems using cisterns in semi-arid Paraiba.* 8º Brazilian Symposium on Rainwater Harvesting and Management. ABCMAC. Campina Grande - PB. 2012.
http://www.bibliotekevirtual.org/simposios/8SBCMA C/8sbcmac-a044.pdf

COLAVIDAS, Felipe; OTEIZA, Ignacio; SALAS, Julián. *HACIA UNA MANUALISTICA UNIVERSAL DE HABITABILIDAD BÁSICA Catalogue of very low-*

cost components, services and installations. Editorial Mareia 2006.
http://www.aq.upm.es/habitabilidadbasica/manuales.html

WHITEHEAD, V. *Development and selection of low cost handpumps for domestic rainwater water tanks in E. Africa.* University of Warwick. 2001.
http://docs.watsan.net/Downloaded Files/PDF/Whitehead-2001 - Development.pdf

MALVEZZI, Roberto. *Semi-arid - a holistic vision.* Brasília: Confea, 2007.
http://www.mda.gov.br/portal/saf7arquivos/view/ater/livros/Semi- %C3 %81 rido uma vis%C3 %A3 o hol%C3 % ADstica.pdf

DODERLIN DE WIN, Theodorus Augustinus; CAVALCANTE DA SILVA, José. *MARBLE PUMP.* 3º Brazilian Symposium on Rainwater Harvesting and Management. ABCMAC. Campina Grande - PB. 2001.
http://www.abcmac.org.br/fíles/simposio/3simp theodorus bomb adeboladegude.pdf

Progress on Drinking Water and Sanitation: 2012 Update 1. WHO/UNICEF Joint Monitoring Programme for Water Supply and Sanitation.
http://www.unicef.org/media/fíles/JMPreport2012.pdf

ADAMS, Jhon... [et al]. *Standards on water, sanitation*

and hygiene for schools in contexts of scarce resources. World Health Organisation, 2010.
http://whqlibdoc.who.int/publications/2010/97892435 47794 spa.pdf

ANAYA GARDUNO, Manuel. *Rainwater harvesting systems for domestic use in Latin America and the Caribbean. Technical Manual.* Mexico 1998. IICA Technical Co-operation Agency.
http://repiica.iica.int/docs/B 1218E/B1218E.PDF

SILVA, Milton Nogueira da... [et al]. *Water and climate change: social technologies and community action.* Belo Horizonte: CEDEFES. Banco do Brasil Foundation, 2012.
http://www.cpa.unicamp.br/alcscens/publications/ LivroAguaeMu dancasClimaticas .pdf

MANCEBO PIQUERAS, José Antonio; JIMÉNEZ, Alejandro. *Appropriate technologies to fulfil the Human Right to Water. Hand pumps.* Tiempo de paz (98). pp. 92-96. 2010 ISSN 0212-8926.
http://oa.upm.es/8726/

ANDRADE NETO, C. O. *Sanitary safety of rural cistern water.* 4º Brazilian Symposium on Rainwater Harvesting and Management. ABCMAC. Juazeiro, PB. 2003.
http://www.abcmac.org.br/files/simposio/4simp cicero

segurancasanitaria sdaaguadecistema.pdf

ALBERTINA DE FARIAS, Silva... [et al]. *MANAGEMENT AND CONSERVATION OF RAINWATER HARVESTING AND STORAGE SYSTEMS IN THE SERTÃO AND CARIRI REGIONS OF PARAIBANO.* 8º Brazilian Symposium on Rainwater Harvesting and Management. ABCMAC. Campina Grande - PB. 2012. http://www.bibliotekevirtual.org/simposios/8SBCMAC/8sbcmac-a036.pdf

TAVARES, A. C... [et ã\]. *Rainwater harvesting and management in cisterns: a way to mitigate the effects of prolonged droughts in the semi-arid Northeast - Case Study: Paus Brancos Settlement, Paraíba.* 6º Brazilian Symposium on Rainwater Harvesting and Management. ABCMAC. Belo Horizonte- MG. 2007. http://plataforma.redesan.uffgs.br/biblioteca/mostrar bib.php?COD ARQ UIVO=1Q597

ACTION AGAINST FAMINE. *Water, sanitation and hygiene for populations at risk.* France: HERMANN ÉDITEURS DES SCIENCES ET DES ARTS. 2005 .ISBN 2 7056 6499 8

UN. *64/292. The human right to water and sanitation.* United Nations General Assembly, Sixty-fourth session, theme 48 of the programme. 2010. http://www.un.org/ga/search/view

doc.asp?symbol=A/RES/64/292&Lang =S

Manual for implementing the Cisterns Programme. First water. Water for drinking and cooking. Operationalisation of the Programme and Guidelines for Applicants. Ministry of Social Development and Fight against Hunger, National Secretariat for Food and Nutritional Security, Government of Brazil. Brasilia 2011.
http ://www.mds. gov.br/segurancaalimentar/programa-cistemas/entenda-o- programa/manual-de-identidade-visual-do-programa-cistemas

I want morebooks!

Buy your books fast and straightforward online - at one of world's fastest growing online book stores! Environmentally sound due to Print-on-Demand technologies.

Buy your books online at
www.morebooks.shop

Kaufen Sie Ihre Bücher schnell und unkompliziert online – auf einer der am schnellsten wachsenden Buchhandelsplattformen weltweit! Dank Print-On-Demand umwelt- und ressourcenschonend produziert.

Bücher schneller online kaufen
www.morebooks.shop

Printed by Books on Demand GmbH, Norderstedt / Germany